KINDLE FIRE

10 Reasons to Get the New Kindle Fire ASAP and Enjoy Your Kindle Devices

By JOSHUA ELANS

Table of Contents

Introduction

Want a tablet that is far better than iPad Air? Then this book is for you!

As the name suggests, you can find out in this Book answers to foregoing questions:

> ➢ Why do you need to buy the New Kindle fire ASAP?

> ➢ Why is Kindle fire better than iPad Air?

> ➢ How can you get started with your Fire Kindle Tablet?

> ➢ What support can you get for your tablet?

> ➢ **PLUS *Advanced Kindle Fire Tablet Tips and Tricks***

It is expected that by the end of your reading, or even halfway through it, you have already purchased your own Kindle fire and had a hands-on experience with the latest in the Amazon Fire HDX 8.9 Tablet.

Amazon's Kindle fire 8.9 is the latest among the Amazon's tablets family and considered a necessity among the self-published writers community. With all its features designed to complement a writer and reader's needs, Amazon indeed managed to come out with a reader that encompassed all traditions when it breaks the tablet market by coming up with affordable high-quality tablets.

With this advanced technology, you will be combining a new gratifying experience of reading, viewing, listening, playing games and much more. May this Book will guide you through those amazing experiences and made everything quick and easy for you.

Chapter 1: Overview on Ebook Readers

The technology for eBook readers had emerged and is now in the mass market. Its demand is too huge it's almost tripling every year due to its popularity since 2006. This is largely due to the widespread trend of eBooks reading and the overflowing supply of digital books. Amazon had set the trend when it opened the door for self-publication in its effort to sell their Prime Program. Amazon's Prime is a marketing effort which provides members in return for their $99 annual membership fee unlimited access to music streaming, to free books in Kindle libraries plus 2 days of free shipping.

With the self-publishing opportunity provided by Amazon, it was able to meet the need of readers and writers to their advantage. They were helping these writers exposed their books to the world, but Amazon got their profit out of revenues from books sales, registration of Prime Members plus the sale of their own Fire Kindle devices.

Before Amazon take its dip into the tablet market, the introduction of the iPad had a big impact on the market for eBook readers in one respect. However, Amazon has driven the costs of these readers and was able to bring down the prices of their own eBook reader family very quickly. The Kindle fire readers are far less than the $499+ iPad 2 tablets, with some models reaching as low as $49.99. The threat from the iPad had forced Amazon to consider launching its own eBook reader.

Chapter 2: Meeting the Kindle fire

The Kindle fire 8.9 is probably one of the most reasonably-priced e-reader available. While the Kindle fire of Amazon is not iPad 2, it definitely offers great value beyond its price.

The Body

Kindle fire 8.9 is designed for convenience and comfort. The body is thin and light that it can be held comfortably with one hand. However, its body is built from magnesium with a blend of glass and nylon, resulting to a durable tablet that weighs 13.7 ounces for 4G version. As it is engineered for increased usability and designed with clean lines, small bezel, and optimized button and charger port replacements.

The Screen Resolution

Amazon goes all out with the new Fire HDX 8.9 as it has the highest resolution (2560 x 1600) and the best pixel density (339 ppi), making nearly a million more pixels than iPad Air 2 for superior viewing experience. It displays vivid and lifelike images that go beyond HD resolution.

The display feature has a dynamic color contrast which automatically optimizes the colors and enhances the viewing experience regardless of the surroundings. It provides clear and colorful viewing in both indoor and outdoor as it features a perfect 100% sRGB color.

Lastly, it features a Dynamic Light Control which changes the white point of the display to match the surrounding in order to resemble a paper-like appearance, making e-reading a comfortable experience in a tablet.

The Processor

Sporting an ultrafast 2.5 GHz quad-core processor and totally delivering quick app launches and smooth multitasking, up to 4x Wi-Fi, and an exceptional Dolby Atmos, HDX can easily outperform other e-readers in providing users with an amazing experience in surfing, video watching, listening to music, and reading eBooks.

The Adreno 420 graphic processor likewise delivers console-quality graphics in high frame rates. Its graphic-intensive games and advanced apps are ultra-responsive and run smoothly while requiring less battery power.

Dual HD Camera

The HDX device is installed with a dual HD camera on its front and rear (8 MP high resolutions) designed to capture superb-looking photos and shoot stunning 1080p video while it also connects you to friends and family with HD video calls.

It also comes handy with its unlimited photo storage from Amazon for all your Amazon contents and photos taken from your device. Editing and sharing of photos are easy and attainable by simply tapping on your images and sending them via email.

Fire OS 4

Powered by OS 4 Sangria, Fire HDX introduces an updated visual designs and platform updates and includes new exclusive feature including Profiles, Family Library, Smart Suspend, etc. It is not just a simple device as it enhances and integrates android technology into the software device, content, and cloud for a simple and seamless experience.

With its personalized profile, family members can now share one device and still have their own personalized segment including individual home screen, email, and social media accounts. This also includes a family library for easy reading and content-sharing.

Incredible Battery Life

HDX OS 4 improves the battery life of your device through the Smart Suspend feature which automatically turns wireless off, thereby saving battery life.

Chapter 3: SOS Amazon!

While chat solutions became a regular part of a support system, Amazon manages to create a support service system exclusively available and unique on Kindle HD products. This system is called "Mayday".

Mayday is a one-click solution that allows users to work with a remote technical representative in your screen. It's like personally working together with just a screen away from each other. This is in line with Amazon's 15-second response time goal when assisting users.

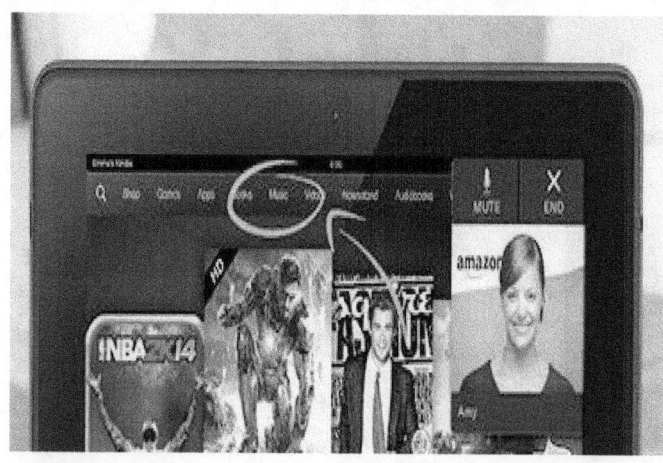

With just a tap on the **Mayday** button from **Quick Actions**, you can easily connect to an Amazon Technical Support Team on your Fire HDX. This technical representative will then guide you through any area on your device by drawing you to your screen, instructing you to either do it yourself or let them do the job for you. Mayday is available 24/7 free of any charge. You will be seeing your technician clearly on your screen but he won't be seeing you if you prefer to shut off your camera.

To activate your Mayday connection, you need to be registered on an Amazon account and have a strong active Wi-Fi connection. To be able to set up your Mayday connection, follow the steps below:

1. From the top of the screen, swipe down and tap the Mayday icon.
2. From the Mayday screen, to enable Mayday, tap **Connect**.
3. If you want to disable the Mayday activation, swipe down from the top of the screen and then tap the **Mayday**.
4. From the left edge of the screen, tap **Settings**, then next to **Mayday,** tap **off**.

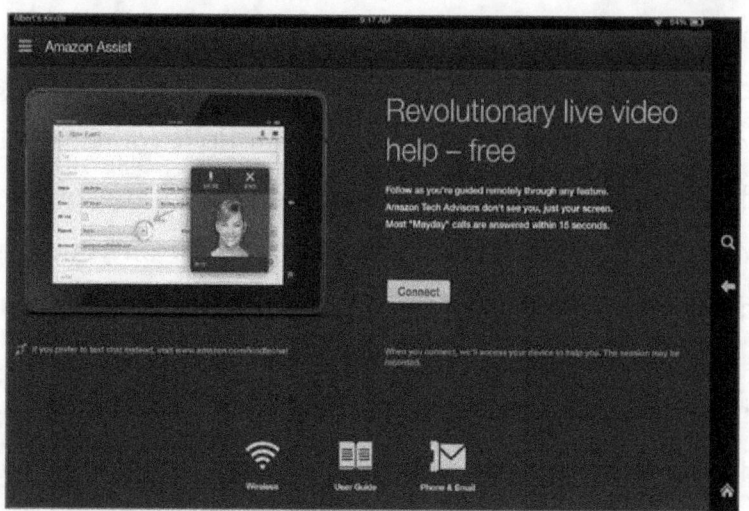

Chapter 4: 10 Reasons Why You Need to Get a Kindle fire

Kindle fire has become a necessity for readers and writers in the Amazon world with its entire feature complement basic publishing tools. However, even if you are not a writer or an avid reader, you have every reason to get hold of the new Kindle fire 8.9 tablet. Here are 10 reasons that will surely sweep you off your feet.

Unlimited Music Streaming and TV Viewing

Amazon Kindle fire and the rest of the Fire tablet family are integrated with Amazon's content and services. One noteworthy service is that Amazon's Prime subscribers are allowed unlimited music streaming and offline TV viewing for a yearly membership of $99. But that is reasonable enough when for less than $9/month, one can have free shipping on items bought from Amazon, access to your favorite TV episodes, and for a specified period, you can avail of unlimited free eBooks and access to Kindle libraries.

#1 - Amazon X-Ray for Movies and TV

Many publishing platforms across the net are offering a wide array of books, movies, and TV shows. Though their offers are usually comparable as each had their own pros and cons, Amazon has a unique feature that gave it an edge over the other competitors. Amazon offers X-Ray for movies and TV, a feature that provides you with a quick gives you a view of the content you want to view.

Powered by IMDb, the X-ray for instant video allows you to see the cast members in a specific scene, learn about the soundtrack as well as the artist, read trivia questions, check out random facts about the film or video and see location information.

#2 – High-Resolution Display

The vividness and clarity of the Kindle fire make it suitable for fine viewing, either for videos or reading. Kindle fire 8.9's Display resolution is higher than that of iPad Air and suitable for watching HD movies and videos. The screen is vividly sharp and amazingly colorful, bringing true-to-life images while adjusting itself to outdoor and indoor surroundings.

#3 - Amazon Mayday Tech Support

This is a state of the art revolution in technology. Never did it happen in the history that a tablet was provided a remote virtual assistance for your technical needs. However, this time, through the Mayday button, Amazon attached this feature side by side with their Kindle fire. If you're the type who would rather not talk to a live representative, you may use the chat option instead and send this video cam off.

#4 - Kindle Fire Tablet Uses Regular Power Port

With the Fire HDX 8.9, you will never have to worry using any standard adaptors or charger, as it does not use any exclusive power accessories. You can always make use of any available chargers and accessories as long as they are compatible with your Fire tablet.

#5 – Backup by a Simply Amazing OS

Fire HDX 8.9 is backed up by customized version of Android called the Fire OS. If you already used to have Android, this will provide you a smooth transition. However, Amazon, make sure that the Fire OS is totally unique from any other versions of Android. For one, it is not possible for Fire OS to gain access to Google Play Store for apps, music videos and other contents. Instead, it uses Amazon's own library and Apps Store.

#6 – Comes with Feature-packed Apps

Amazon always offers mobile apps for almost all of their mobile platforms such as iOS, Android and Windows Phone (Kindle Only) and you can expect more polished and feature-packed apps from your Kindle fire.

- Microphone
- Front-Facing Camera
- 3.5mm Stereo Jack
- Dual Stereo Speakers
- Rear-Facing Camera
- Flash
- Volume Buttons
- Power Button
- Micro-B USB Port

A Good example is the Goodreads Social Networking App for readers which is directly integrated into the kindle application. This provides you easy access to connections and view as well as share books with them. Goodreads is available for iOS and Android but is not integrated with the Kindle apps for these platforms. Kindle app for Fire HDX has also much more advanced viewing and notation features.

#7 –Ultrafast Wireless Connection

With you new Wi-Fi HDX tablet, you can experience a full-speed connection regardless of what Wi-Fi network service you have.

Through the Fire HDX 8.9, Amazon displays the latest mobile wireless technology. The integration of MIMO method with the 802.11ac mobile wireless technology makes your tablet connection with a capacity of optimizing internet's bandwidths of up to 600 Mbps. It is said to quadruple the connection speed of the previous-generation Fire HDX.

It enables you connect to the internet, download and stream at ultrafast speed. For maximum performance, Fire HDx 8.9 operates on multiple bands to ensure the fastest possible connections.

#8 - Amazon Instant Video Web Interface

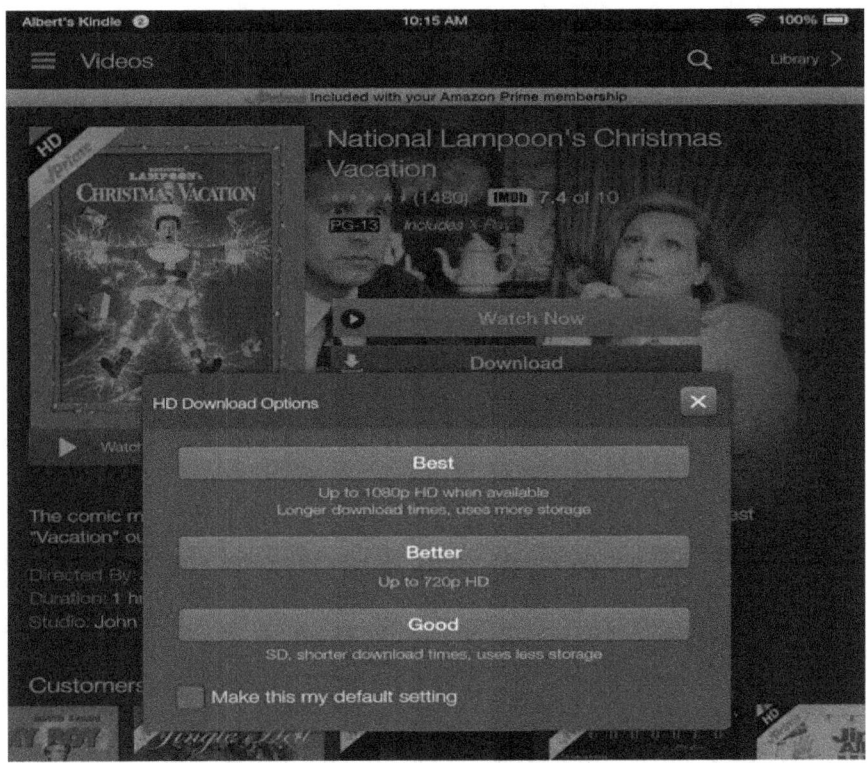

All those digital contents that you can buy from Amazon such as books, music, movies and TV shows can be accessed via a browser. Hence, you don't need to be using an Amazon device to access your purchases. This is one thing that is more convenient with Amazon. You can readily download a movie from Amazon Instant Video or music via the Amazon MP3 Store and you can do all these by simply using major browsers. Accessing videos and other contents from Apple are not possible other than using Apple devices. You can never download these contents since there is no iTunes application in non-Apple devices.

9 – Watching on the Big Screen

There can be nothing more exciting than watching movies on the big screen. With all those files of movies and television shows uploaded on your tablet, you can watch them like watching on a big screen. You can now make use of your TV set as Second Screen and watch movies just like you're watching them on a real big screen.

#10 – Firefly technology for Music, Movies and Television

The firefly app instantly recognizes more than 240,000 movies, songs, and TV episodes so you can access artist's information.

Immediate Video Starts with ASAP

This time, you don't need to waste time waiting for videos to buffer. ASAP, short for Advanced Streaming and Prediction anticipates what movies and videos you want to watch and make everything ready for you even before you hit the play button.

Chapter 5: Getting Started with Your New Kindle fire

Having your own Kindle fire for the first time can leave you hanging in mid-air especially when there's no one there to guide you in setting it up. Of course, there is a manual included in your pack, but manuals are not interactive and sometimes hard to grasp.

In this section, you will be treated to a thorough understanding of how you will manage your device and you will also be provided with a step-by-step instruction on how to do it.

Through this guide, you will be shown the proper way to set up your new Amazon Fire HDX 8.9 tablet. We will be walking with you in every step to get your tablet up and running and the best thing to do even after you get through with the basics as you turn your tablet on for the first time.

This guide can work even for the new Fire HD 7, Amazon Fire HDX and Amazon Fire HDX 8.9. This will help you get your Amazon Fire tablet running on the right track.

Startup Procedure

As soon as you receive your tablet and get it out of the package, you look for the power button to set it on. Then adjust the power volume and alter the screen rotation setting. Activity with your tablet can be done either through touchscreen interaction or by pressing the hardware buttons.

Turning on the Power – You can turn your Fire HDX power on by just pressing and holding the power button located on the underside of the tablet for 2-3 seconds. (Note: In case you want to restart your tablet, just press and hold the Power button for approximately 2-3 minutes and then you will see the notification; "Do you want to shut down...?" Select the right button to turn it off. When the device has completely turned off, it's time for you to press the power button to restart it.

- **Adjusting the Volume** – At the back of your device, you fill find the + or – button, press any of these two buttons to increase (+) the volume or decrease (-) it.
- **Rotating the Screen** – You can rotate your tablet horizontally to get into the Landscape mode, or vertically for the Portrait mode.

Charging the Battery

You may not need to charge the battery while initiating the above procedure. However, if the battery symbol shows that the battery load is below half, then you need to recharge it before you can start setting it up and registering your device.

Hardware Basics

Charging the Battery – Your devices is equipped with a micro-USB and adapter. These are included when you buy your device. If you use other power adapter or USB cable, chances are that your charging time is increased.

To charge, simply connect the micro-USB cable to your Fire tablet and also to the power adapter. Plug the power adapter into the power outlet and leave it to charge until full. Note that in the Kindle fire device, the micro-USB cable never appears straight but slanted when connected to the USB port. This is to allow you comfort and convenience while holding your device when it is charging.

When your device is charging, you will find a lightning bolt in the battery indicator.

Connecting to Wi-Fi - Your device is attuned to automatically detect Wi-Fi networks operating nearby and wireless hotspots that broadcast their network names. There are some networks that are open for public while others are secured with passwords. You can connect your device to your own home Wi-Fi network if you have one. Here's how to connect.

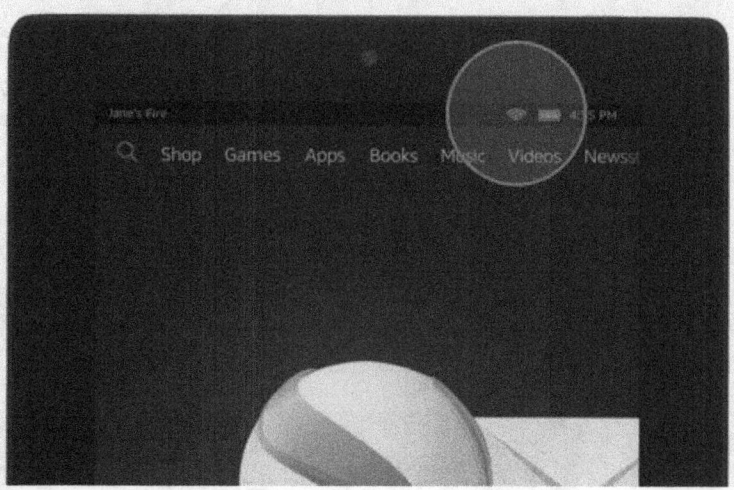

1. From the top of the screen, swipe down and tap **Wireless.** Make sure that your device is not set to **Airplane Mode.**

2. Press **Wi-Fi**.

3. Next to Wi-Fi, Tap **On**.

4. Choose a network to connect. If you see a lock icon, it means that the network is secured and a password is required. Enter the password and tap **Connect**. In case you don't know the password, you may ask from the person who set it up. Note that your Amazon password is different from the password which is used in your wireless network.

Right after you have connected to your Wi-Fi Network, your device will connect automatically to the network when it is in range. If there is more than one network within a range, your device will automatically connect to the last network used.

Registering your Fire Tablet

For buying content and delivering it to your device, you need to register your Fire HDX device. This also allows you to transfer purchases across devices and readings apps. Here are the steps involved in the registration.

1. Starting from Home page, swipe from the top downward and tap Settings.
2. Follow by tapping **My Account.** Then tap **Register**. If you already have an Amazon account, just provide your account information and then tap **Register.** However, if you don't have an existing account, then set up a new account by following the instruction.
3. In case you want to opt out of Amazon, you may deregister your account by doing this through **Setting** in **My Account**. Tap **Deregister** and you're out of there.
4. In your computer, go to **Manage Your Content and Devices**. Choose **Your Devices**.

5. Select your device or app and then again select **Deregister** from the Action Column.

6. Confirm your action by selecting **Deregister** from the pop-up window.

7. You may register to another Amazon account or if you are selling your device, restore it to factory setting. All downloaded items then will be removed from your device.

Fire Tablet Basic

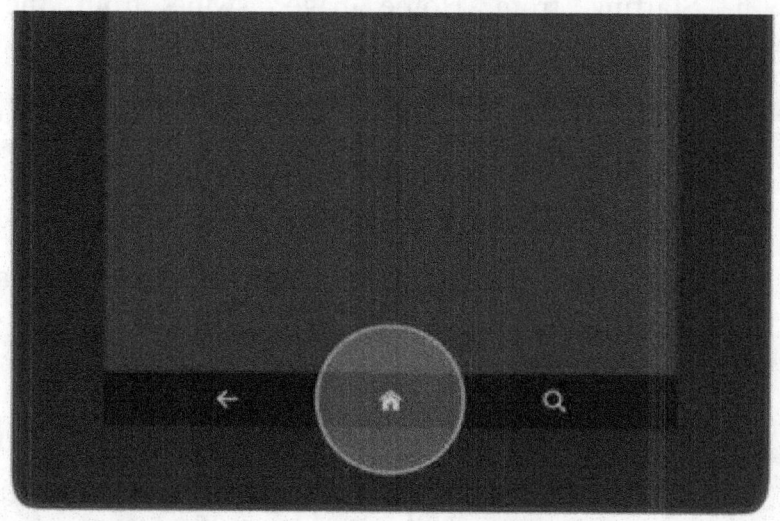

Navigating Your Tablet

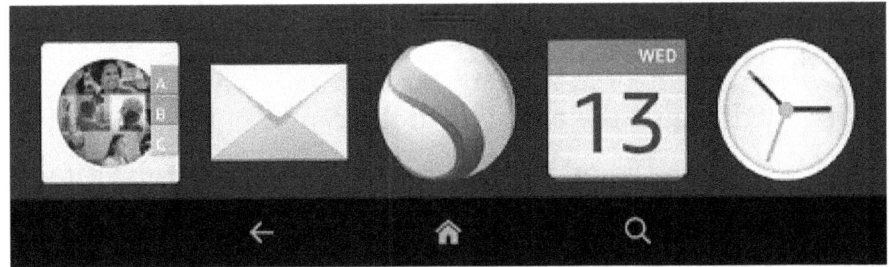

- To access the options bar while reading, tap the center of the screen.
- To access the options bar while viewing content in full-screen mode, swiftly swipe down the screen from the top.
- To go to Home, tap the home icon.
- To go to the previous screen, tap the back arrow.
- To search for files or apps on the device, digital stores, or online, tap the magnifying glass icon.
- To close the onscreen keyboard, tap the keyboard icon. This appears only if the onscreen keyboard is open.

Accessing Recent Content with Quick Search

You can easily switch between recent contents using **Quick Search**. Just drag your finger from the Options Bar and tap on the item you want to select. As you are viewing an app or web page in full-screen mode, gently swipe down from the upper edge of your screen or from the right to show the Options Bar. From there, drag it to the center of the screen to see the **Quick Switch.**

Tip: By dragging the app icon to the center of the screen, you may also close the applications from **Quick Switch** to force the app to stop. This action will also remove the app from the Carousel.

Managing Sounds and Notification

You may also view and modify notifications from Quick Actions or mute notification using **Quiet Time** in your Fire Tablet. Note that once you plug in your device, it will alert you when it's fully charged. The same thing happens when you press and hold the power button to turn it off. All these notifications can't be turned off.

- **Viewing Notifications**

    1. Open the Quick Actions by swiping down from the top of the screen.

    2. As your notification will appear just below the Quick Actions menu, tap the notification to interact or swipe to dismiss it.

    3. To be able to dismiss multiple notifications, simply tap **Clear All**. You can also view and dismiss notifications from a lock screen. Swipe down to view, access, or dismiss the notification.

4. If you want to turn off this feature, swipe down from the top, tap Settings, and then tap **Security & Privacy**.

5. Just next to **Lock Screen Notifications**, tap **Off**.

- **Modifying Setting for Notification**

To quickly manage settings for that application,

1. Press and hold the notification. Swipe down and then tap **Settings**.

2. Tap **Notifications** and **Quiet Time** and choose an app from the list. This will allow the app to show a notification or give an alerting sound when it arrives for that app.

Tip: If you want to mute all notification alerts and hide notification, just turn on the **Quiet Time** from the **Quick Actions.** You can schedule a specific time to automatically turn on during a certain period like when you feel like reading a book or listen to music.

Accessing Content

Your Fire HDX can store an indefinite number of content like books, music, apps and much more. You can access, organize, and remove this content from your device by connecting to a wireless network and sync or download your content from the Cloud to your Fire Tablet.

Syncing your Fire Tablet

To sync your Fire HDX tablet:

1. Make a downward drag on your screen and then tap **Settings**.
2. Tap **Sync and Check for New Items**.

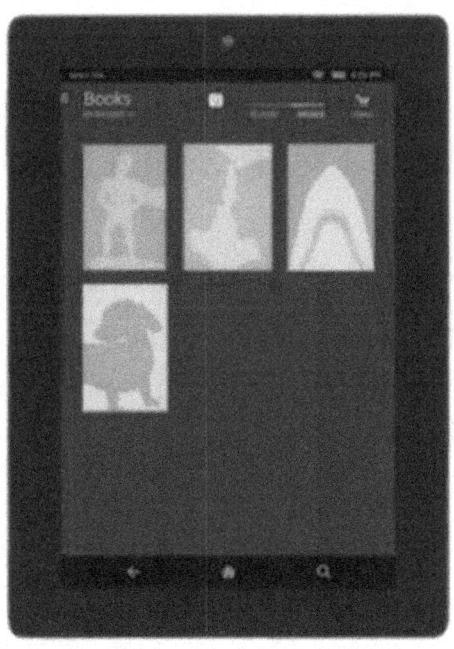

Download content from the Cloud

Every time you purchase content in your Fire tablet, it is saved to the Cloud and you can download your content from the Cloud to your tablet when you connect to a wireless network. To get access to your content,

1. From Home, tap a content library like **Books** in the top navigation bar.
2. Tap on the title to download it to your device. Then after the content has been downloaded, search for the marking in the bottom right corner Items that are still stored in the Cloud don't have a check on it.
3. Tap the title to open it.

Linking Your Tablet to Social Networks

You can connect to social networks by linking your tablet to enable you to share your reading status, book highlights, and ratings. You can also import calendars, photos, and contacts to your device.

1. Swipe down on your screen and tap **Settings.**
2. Tap **My Account**.
3. Tap **Social Networks** and choose from the following: **Connect Your Facebook Account**

Connect Your Twitter Account

Goodreads

4. Log in to your account and then tap **Done.**

   From there, you can have full access to these social networks and provide interaction to your connections. You may also unlink any of these.

Smart Suspend

You can manage your device's battery power by enabling Smart Suspend feature of your Fire HDX Tablet. This will allow Smart Suspend to disconnect automatically your Wi-Fi connection and other connectivity when not in use to save your battery power. It can also be set according to the time of your preference as to when it will automatically disconnect your Wi-Fi connection.

Turning Off Smart Suspend

1. Swipe from the top of the screen downward and tap **Settings**.

2. Tap **Power Management**.

3. Next to **Smart Suspend**, tap **Off.**

Chapter 6: Connecting Wirelessly

Second Screen on Fire Tablet

For a good viewing experience, you may connect your tablet to the Amazon Video app. Your Television will serve as the primary screen while your tablet will act as a remote for playback controls or a customized display for the x-ray if you want to learn more about the movie or show you are watching. You can even perform other tasks on your tablet while the video continues playing on your TV.

You can use Second Screen to integrate your Fire tablet with the Amazon Video app on your Amazon Fire TV, Fire TV Stick, or PlayStation console.

Your Fire TV device is the primary screen. Your Fire tablet acts as a remote for playback controls. It can likewise operate as a customized display for X-Ray featured app so you can learn more about the movie or TV show you're watching. You can also opt to get away from the Second Screen experience so you can perform other tasks on your tablet leaving the video continuously playing on your TV.

Second Screen can be connected to the following streaming media devices

- Amazon Fire TV
- PlayStation 3 (PS3)
- Fire TV Stick

Here's how to connect:

Before you start, make sure that the two devices that you are going to use are powered on and connected to the internet. They don't need to be connected to the same wireless connection.

1. Open the video details on your tablet by tapping Videos.
2. Next, tap the **Second Screen** icon on the **Watch Now** button.
3. When a pop-up window opens, select a Fire TV device to send the movie or TV show to.
4. You will see that your selected video or TV show will simultaneously be shown on your tablet and TV.
5. While watching, you can control playback on your tablet. Tap **Play, Pause, or Jump Back** buttons. You may also use the progress bar to move backward or forward.

6. When the movie includes X-Ray, simply tap **In Scene** to see information relative to the characters, actors, trivia and music. When you're done watching, tap **Second Screen** and select your tablet to end playback.

Display Mirroring on Fire Tablet

You can set both your tablet and compatible TV or media streaming devices to wirelessly "mirror" each other. Through this feature, the two devices will be synchronized to give you a unique video viewing experience.

Before doing this, turn on the two devices - TV and Fire HDX 8.9 tablet - and make sure they are discoverable. The term "Discoverable" here means they are compatible and can recognize the presence of each other. If they are not, you may use HHDMI display dongle, an HDMI adapter that you can use to connect to your TV or media streaming device. With an HDMI dongle, you're allowed to wirelessly display content from your tablet to your TV or other media streaming device.

From the top of your tablet's screen, swipe down mildly and tap Settings.
Tap Display & Sounds. Tap Display Mirroring. Then your tablet will try to detect any compatible devices within range.

Tap the name of your media streaming device or TV. This will take about 20 seconds. After allotted time, your device screen will appear on your TV screen.

If you want to stop displaying the screen, swipe down from the top and tap Stop Mirroring.

Chapter 7: Media and Apps

Managing Photos and Personal Videos

To automatically upload your photos or videos to your Cloud Drive automatically, just do the following steps:

1. Swipe your screen from the left.
2. Tap Settings.
3. From among the selection below Auto-Save, click on the switch next to Photos or Videos.
   Tips: Make sure that your camera is clean and free from any obstructions when you take pictures and videos. Also, keep the device steady to avoid motion blur.

Downloading Photos and Personal Videos from Cloud

Once you are connected to a wireless connection, you can download your images, graphics, and personal videos from your Cloud Drive account to your tablet.

- Swipe from the left and then tap **Cloud Drive**. Your files will appear here.

- Select, then press and hold an album, a photo or video inside an album. Tap and download.
- Swipe from the left of your screen and tap **Device** to see your downloaded item. Note that you can also relocate some of your photos and personal videos from your computer to your tablet using a micro-USB cable or you may upload them to Cloud Drive.

Viewing, Adding and Sharing Photos

You can view, add, edit, share or print photos from your Fire tablet.

From Homepage, tap **Photos.** Then you swipe from the left edge of the screen.

To be able to view your photos or videos tap:

- **All** - View personal videos and images or graphics stored on your device or in Cloud Drive.
- **This Day** - View the photos that you took on this date in past years.
- **Camera Roll** - View recent photos and videos taken with the camera.
- **Videos** - View personal videos kept on your Fire HDX device or in Cloud Drive.

- **Cloud Drive** - View photos and videos stored in Cloud Drive.
- **Albums** - Create, organize, view, and share albums of your photos and videos.
- **Device** - View photos and videos downloaded to your Fire tablet or images and videos downloaded from email attachments or the Internet.

To add photos or videos to your library for viewing your device, simply tap **Add Photos**, and then click:

- Mobile Device
- Facebook
- PC or Mac
- Transfer through USB
- To import your photos to your device, follow the on-screen .
- Buy and Download Apps and Games

You can search for and download apps and games on your Fire HDX tablets.

So you can buy games and apps from Amazon App Store, first set up your payment method. Your payment could either be a credit or debit card.

There are some games and apps that offer In-App purchasing. To turn off In-app purchasing feature from your apps library, swipe from the left side of your screen and tap Settings.

Tap in-App Purchasing

Uncheck the box next to Allow In-App Purchasing.

1. From Home, tap **Apps or Games**, and then tap Store.
2. Search for the app or game you want to buy:
3. In the Search field, simply encode the terms connected to your inquiry, and then tap the magnifying glass.
4. For a shortcut search, swipe from the left edge of the screen, and then tap **Browse Categories, Recommended for You, Best Sellers, or New Releases.**
5. Tap the button with the price for paid app or game or tap **FREE** for an absolutely free app or game, and then tap **Get App.**

You can make purchases through Amazon 1-click or coins. After you buy an app, it will automatically download and install to your Fire Tablet.

To view your app or game, tap Open. If a License Agreement appears, just carefully read and understand the agreement. Quickly tap **Agree** to accept the terms and conditions.

Updating Apps and Games

Your purchased app or game will automatically update when the developer offers a new software version for the app. When the app or game requires new permissions, you will be asked to visit the Amazon App Store on your Fire Tablet in order to manually update the application.

To manually update an app,

1. Swipe from the left edge of the screen in the Amazon App Store.
2. Tap **App Updates.**
3. Tap **Update** to update the app, or tap **Update All to download** in your account the latest version for all apps. Tip: By enabling ***Whispersync for Games,*** you are allowed to sync you're the progress of your game for supported games across other Amazon devices. ***Whispersync for Games*** will save your game data in the Amazon Cloud
4. To enable ***Whispersync for Games*** on your Fire Tablet,

1. Tap **Games**, and then swipe from the left edge of the screen.

2. Tap **Settings**, and then tap **ON** next *to Whispersync for Games.*

Chapter 8: Security and Settings

By opening Quick Actions of your Fire HDX, you can connect to a wireless network quickly, adjust screen brightness, and access additional settings.

Quick Actions

Here are some of the interaction activities you can do from the Quick Action menu.
To access this, simply swipe down from the top of the screen.

- **Auto-Rotate** – Lock and Unlock screen rotation
- **Wireless** – Connect to a wireless network, pair a Bluetooth device, and turn on Airplane Mode
- **Brightness** – Adjust the brightness of your screen
- **Quite Time** – Hide notifications and mute.
- **Mayday** – contact a Technical advisor or representative for a more personalized help session.
- **Firefly** – open the Firefly app.
- **Settings** – access additional device settings such as keyboards, accessibility options, parental controls, date and time and more.

Settings Menu

- **My Account** – To register your device, manage accounts, and connects to social networks.

- **Sync and Check for New Items** – To sync your device to download the latest software or sync to the latest page you have read on your Kindle books, and update your apps.

- **Household Profiles** - To enable family members to add profiles for a more personalized experience on a shared tablet.

- **Help** – To get access to Mayday assistance, browse for online help, or contact Amazon assistance or technical support.

- **Device Options** – To view device free storage space, change device name, to provide back up support for your device, alter date and time, install updates, etc.

- **Wireless & VPN** - To connect to Wi-Fi, pair Bluetooth devices, enable Airplane mode and many others.

- **Power Management** – To enable Smart Suspend and manage Display and wireless settings.

- **Applications** – To configure application settings for Amazon applications or third-party applications.

- **Display and Sounds** – To adjust your screen brightness, volume, and settings for Display Mirroring.

- **Language & Keyboards** – To change your device language and settings. You can also set the voice for Text-to-Speech.

- **Notification & Quite Time** - To manage notification sounds for apps and set a schedule for Quiet Time.

- **Accessibility** – This is to assist vision and hearing-impaired users – to manage settings for Screen Reader, Closed Captioning, etc.

- **Parental Controls** – To manage child profiles within your household and to turn parental control off and on.

- **Legal Compliance** - To view privacy policies, legal notices, terms of use and more.

- **Security & Privacy** - To set a lock screen password, enable device encryption, and more.

Chapter 9: Fire Tablet Quick Fixes, Tricks, and Tips

Got your new Kindle fire 8.9 or the rest of the HDX versions? To get the best of your device, here are some handy hints. These will get you fully up to speed on your Kindle Fire tablet.

Side Loading Android Apps

Though Amazon's Kindle Fire Tablets operates on an Android's customized version and had many similarities with Android tablets like Nexus 9, still it is exclusive to Amazon and has no access to Google Play Store. Amazon may have loaded its slates up with Amazon Appstore where it sells its own version of popular apps. However, it's missing big players, notably all of Google's own applications. To side load these apps on your device. Here is the way to do this.

Go to Settings app, then to Applications. Change Apps from Unknown Sources from off to on.

Find the app, which is the package file format used by Android apps. On the device browser, search for www.appsapk.com and head in there. Make a search for the app you want, then tap to download. It will download directly to your device.

Installing Another App Store

If you want to level up this sidelong process of installing apps from outside source, then you may as well install another App Store. It works the same way you did with downloading the app. Only this time, search for the Store App like *SlideMe* and *GetJar*.

So it's just as simple as that!

Taking Screenshots

It's a neat trick to learn taking screenshots on your Amazon Fire tablet. If you want to take a quick snap of your current screen, merely press the volume down and standby buttons at the same time. Your screenshot images will be stored in the gallery app.

Uploading Content to Amazon Cloud

Cloud Storage is quite handy these days and if you own an Amazon device, then you have the best option of the Amazon Cloud Drive. This Dropbox-like service starts with a 5GB free storage and you can expand it to 20GB of space by purchasing additional space for a storage fee. However, if you are an Amazon Prime subscriber, you can have unlimited storage for your photos and music download purchased from Amazon for free.

Anything that you have stored in the Amazon Cloud is accessible through a web portal or directly from your device. Uploading files to the Cloud Drive from your browser on a tablet is easy. Just over to the Homepage of the Amazon Cloud Drive, log on using your Amazon details and tap the (+) icon on the top right corner. Drag the file you want to copy. Now, go and check on your tablet and you will find there the file that you had just uploaded.

Doing More with the Carousel Interface

The carousel interface is exclusive to Amazon tablets. While it's hard to customize on this, there are some tricks you need to learn. If you want an icon removed, simply press on it longer and it will be removed from the carousel though it still exists in your device. You may add it to a Collection. A Collection is a folder you can use to organize your content.

Some Quick Fixing

There may be some troubles with your Amazon Fire HDX that you want to resolve on your own. Try learning some quick fixes.

Troubles	Quick Fixes Guide
If your screen is unresponsive	1. Restart your device by holding the button for at least 40 seconds until it restarts automatically, then release. If it fails to restart automatically, press the power button.

	It must lead you to the startup screen immediately.
If the Screen is slow to start.	Consider the following: • Your version of the app needs to be updated. Visit Fire & Kindle Updates. • You may be currently downloading something. • Do not use your device under extreme temperature. • Touchscreen is dirty. • You have a protective case or screen. Try removing these accessories and restart your device.

Forget Lockscreen Password or PIN	
	1. On your device, go to **manage your Content and Devices.**
	2. If you have more than one device listed on Amazon, select the device that needs to be fixed.
	3. Make a click on **Device Action** from the drop-down menu. Select **Remote Lock**.
	4. From your Fire HDX, tap **Unlock Device**.
	5. In unlocking your device, just make a swipe from the right edge of the screen to unlock your device.
	6. Enter your screen password or PIN
	7. Tap **OK**.

Conclusion

By now, you are done with your reading and had started some hands on with your Fire Kindle HDX, you probably had mastered the basics, learn some quick shortcuts and quick fixing.

With its specs designed and built for maximum performance at a price, you would have less expect it, buying this astounding device can be a necessity. Everything in it makes it a worthy rival of Apple and though it carries a number of features that can outperform iPad Air, each has their own share of usability and performance. But for something like its price, Kindle HDX 8.9 proves to be a winner!

Hoping this eBook had thoroughly guided you through every step of setting up your astounding device. But if until now you have not yet purchased it, then better hurry up. Buy yourself the new Amazon Kindle fire ASAP!

www.ingramcontent.com/pod-product-compliance
Lightning Source LLC
Chambersburg PA
CBHW070403190526
45169CB00003B/1082